Edwin Webb

Memorial sermons

The capture of Richmond

Edwin Webb

Memorial sermons
The capture of Richmond

ISBN/EAN: 9783337112950

Printed in Europe, USA, Canada, Australia, Japan

Cover: Foto ©berggeist007 / pixelio.de

More available books at **www.hansebooks.com**

MEMORIAL SERMONS.

THE CAPTURE OF RICHMOND.

SOME OF THE RESULTS OF THE WAR.

THE ASSASSINATION OF THE PRESIDENT.

BY

EDWIN B. WEBB,

PASTOR OF SHAWMUT CHURCH, BOSTON.

◄•►

BOSTON:
PRESS OF GEO. C. RAND & AVERY, 3 CORNHILL.
1865.

REV. E. B. WEBB.

Dear Sir,—

The undersigned beg leave to ask a copy of each of the discourses delivered by you in the Shawmut Church on the mornings of the 9th, 13th, and 16th inst., for the press.

Approving the sentiments which they contain, and believing them adapted to perpetuate the memory of those *momentous events* which followed in such rapid succession, and which gave to a great nation so much joy and such intense sorrow, we hope you will not deny our request.

Very respectfully yours,

L. R. SHELDON,
DANIEL HARWOOD,
OSBORNE HOWES,
CHARLES HUTCHINGS,
C. A. PUTNAM,
MOSES W. RICHARDSON,

Committee.

BOSTON, April 20, 1865.

GENTLEMEN,—

The sermons for which you so kindly ask are very humble and hasty efforts; but, indorsed by a large and intelligent congregation of citizens, they may help to shape the popular opinion, and to record the fearful struggle between loyalty and treason, and also the sudden transition from one extreme of unutterable feeling to another, through which we have passed in these times which have no parallel in national history. They are, therefore, submitted to your disposal.

With ever-increasing affection,

I am faithfully your pastor,

E. B. WEBB.

To L. R. SHELDON, M. D., DANIEL HARWOOD, OSBORNE HOWES,
CHARLES HUTCHINGS, C. A. PUTNAM, MOSES W. RICHARDSON, Esqrs.,

Committee.

THE CAPTURE OF RICHMOND.

> " The Lord is my strength and song, and he is become my salvation : he is my God,
> and I will prepare him a habitation; my fathers' God, and I will exalt him."—
> Exodus xv. 2.

THE past has been a week of great joy. Not one
city, not one State, but the whole loyal North, from
Maine to California, has been jubilant. Not waiting
for official requisitions, the people — spontaneously,
unitedly, universally — have given thanks to God in
prayer, have united in songs of thanksgiving, have
thrown their flags and streamers to the winds, have
made their houses from attic to cellar luminous with
expressions of gladness; and at the call of authority,
citizens have assembled with bands of music, amid
the peal of cannon, with waving banners, in vast
crowds, to congratulate each other, and give utterance
to the general feeling of exultation.

And manifestly this has not been a mechanical
self-excitation: men have been moved; hearts have
been oppressed with feelings that words could not
utter; and the shouts of living, sympathetic masses
have been called for, and Te Deums with full orchestra.

to express even partially what every one has felt. We have rejoiced before, but then with fear and dark forebodings; now with no vague, oppressive apprehensions, with no reservations, but with full and deep respiration, and unclouded hope. Men feel that the foundations are sure, that the instincts of a loyal people are right, that justice lives, that God reigns.

And what is the occasion of all this gladness and congratulation? Just that which called forth the Song of Moses from which our text is taken: "Thy right hand, O Lord, is become glorious in power: thy right hand, O Lord, hath dashed in pieces the enemy."

A little more than a year ago, a modest, reticent, but able and magnanimous commander was called to Washington, and made commander-in-chief of all our armies. Up to that time, the strength of the Rebellion had been once or twice checked, but never humbled. The leader of those rebel legions had never met his master. He was still abroad, free, and able to move when and where he pleased. When he chose, he fortified and lay behind his works; when he chose, he invested the capital or invaded the loyal States. His power was unbroken, his spirit undaunted, his limbs unfettered.

With a quiet determination, our new commander said that *General Lee must be controlled*, his army must be conquered. The strength of the Rebellion is here. Operations aggressive and successful

elsewhere are worthless while the heart of the
Confederacy is untouched. Quietly but efficiently,
as the forces in Nature accomplish their end,
General Grant set himself to his unmatched and
terrible work. The sight is sublime, as, with his
carpet bag in his hand, and a single attendant at
his side, he crossed the Potomac to undertake and
execute his masterful design. On the first secular
day of *May*, 1864, he crossed the Rapidan to find
and fight the enemy, — the enemy of the Republic and the enemy of mankind. What work
followed, for days and nights and weeks and
months, in the Wilderness, at Spottsylvania, and all
the way down to Cold Harbor, across the Pamun-
key, and finally across the James, and to the en-
virons of Richmond and Petersburg, you all know.
How many brave fellows found a bloody grave;
how many thousands found the hospitals only to
sleep there their final sleep; how many more
thousands found themselves sent home with hon-
orable wounds, — will be known wherever history
is read. Such fighting, so determined, prolonged,
and persistent; such courage; such fortitude; such
marching; such endurance of heat and wet, inces-
sant fatigues and bloody battles, — the world never
saw.

And how much we owe to that tenacity of purpose which never let go its hold, and never was

diverted from its aim, which fought till the blood stood in pools on the ground, and never failed to follow up an advantage, — is not known yet, as we believe. It evinces the true greatness of the Lieutenant-general, to sit there before Petersburg in grim patience, planning brilliant campaigns for popular leaders, and praising them with shotted salutes for their splendid execution; doing nothing himself that the world can see or praise or photograph. There is something of moral heroism, as well as of intellectual greatness, in the man who thus silently holds the organized military strength of the Rebellion with the powerful grasp of one hand, and with the other points out what can be done in the Valley of the Shenandoah, and at Nashville, and along the Atlantic coast, and through the central and vital parts of the Confederacy.

After a council of war, on Monday night, March 27, 1865, at which were present men whose names will head new chapters in history, — Lincoln, Grant, Sherman, and others, — the great army of the Lieutenant-general began to move, under his immediate supervision. The solution of the mighty problem which has engrossed not only the thoughts and energies of this nation, but has interested and influenced to a great extent the nations of the whole earth, is about to be solved. A bold, stupendous plan is just unfold-

9

ing its ultimate design. For five long, weary days, marching through dense forests, floundering through swamps, fording streams, drenched with rain, exhausted in the mud, catching a few moments' sleep in the ranks, or upon the cold spongy ground, while waiting for orders to move, — for five weary, memorable days the uncertain struggle has raged. Back and forth over the same ground the veteran combatants, horse and infantry, swayed and surged. Friday closed gloomily for the loyal cause. The rebel leaders had become fully aware of the disposition of our forces, and of their own immediate peril. And, putting forth all their long-enduring, disciplined strength, our cavalry was driven back ; our infantry was repulsed ; our veterans of the fifth corps, many of them, were killed, wounded, captured. But, as if helped by the inspiration of the Almighty, and confident that the hour and the end were nigh, our lines reformed, and, becoming steady, once more moved forward, like a good ship that feels her helm again after the staggering shock of some tremendous wave. Gaps are filled up, connections made ; and feeling one another's strength, and the confident, steady *will* of a present commander, the aspect of the whole field is gradually changed. An unseen hand ministers courage to fainting hearts. Cheers break out in well-known accents along the Union lines. The crisis is passed. Forward, forward! and *wrong* long defiant, cowers before the *right*. Rebellion acknowledges its master.

And now what means this joy of the loyal people, this exultation of loyal hearts seeking expression through every possible symbol of gladness and congratulation?

It means that the serpent, driven hissing and defiant into her den, and kept there by a cordon of fire, is making her escape with the whole slimy brood, along a tortuous way, — the only one open to them, — dodging and dipping their vile heads to avoid the conqueror's uplifted heel.

It means that victory, complete and final victory, crowns the loyal cause. It means that the fortifications and forts, and all the defences of the rebel stronghold, are stormed and carried. It means that the spirit and strength of Rebellion are at length subdued and conquered. It means that the so-called Confederate Government is a vagabond, without a capital, without a home, without resources, without revenues, without currency, without friends, and execrated by all peoples under heaven. It means that Lee, who was educated by the Government, as by a munificent mother, at her best-endowed schools, and in her best positions; who remained with General Scott just long enough to learn unmistakably all the resources and secrets and purposes of the Government; and who, for the last four years, has been using all

that knowledge, and all his education, and energy,
and military skill, that with perjured hand he might
plunge his sword into the bosom that nursed him, —
. it means that Lee, the pet and pride of the Rebellion,
is foiled and vanquished at last.

The old mother has had too many sons to stand
between her and the weapons poised and aimed by
traitorous hands. Lee is flying, a criminal, he knows
not where, to escape a merited halter, he knows
not how ; his army beaten, his braves captured, his
followers deserting.

It means that the Lieutenant-general and the loyal armies are master of the field, and master of the
Confederacy. It means that slavery, the one dark
cause of all our bitterness and bloodshed, is forever
dead. It means, no more massacres at Fort Pillow ;
no more Union soldiers in Libby Prison, nor starved,
emaciated, dying skeletons at Salisbury or Anderson-
ville, — names to be pronounced with righteous indignation, and remembered with a shudder. It means,
no more threats on the floor of Congress, no more
tyranny of opinion, no more forced silence anywhere,
no more iniquitous compromises, and no more repeal
of righteous ones. It means that the old flag, which
was planted, in the resplendent morning of April 3,
1865, on the ramparts of Richmond, conquered and
captured by Liberty's thundering legions, shall float,
the emblem of strength, freedom, and justice, over

every capital and city and village and inhabitant of
our original domain.

It means that the war, the long terrible war that
has trampled down our harvests, and burdened our
industry, and desolated our homes, and bereaved our
hearts, and made delicate wives widows, and helpless
babes orphans, — this war for the overthrow of a
beneficent government is virtually ended.

It means that Republican Government, a Government of the people and by the people, is no longer
a question and an experiment. It means that the
people have the *right* and the *capacity* to govern
themselves in peace and in war; and, therefore, that
the Republic of the United States of America is a
government and a success. It means that our institutions, affording security and liberty and opportunity and education and religion to *all men*, are to
triumph.

Such is the occasion of our joy to-day; and well
may we congratulate each other, and utter forth
praise, and take up the strains of ancient Israel,
"The Lord is my rock, and my fortress, and my de-
liverer; by thee I have run through a troop, and by
my God have I leaped over a wall." Well may we
repeat after Moses and Miriam, taking up that sublime strain which the shore echoed to the sea, and
the sea back to heaven again, "I will sing unto the
Lord, for he hath triumphed gloriously. . . . Thy right

hand, O Lord, is become glorious in power, thy right
hand, O Lord, hath dashed in pieces the enemy."
" He maketh wars to cease unto the end of the earth ;
he breaketh the bow, and cutteth the spear in sun-
der; he burneth the chariot in the fire."   " If it had
not been the Lord who was on our side when men
rose up against us, then they had swallowed us up
quick when their wrath was kindled against us."

Let me not fail, my friends, to turn your thoughts
and your thanksgivings directly to God.   As of old,
when the prophet warned the king, " Behold the people shall rise up as a great lion, and lift up himself as
a young lion: he shall not lie down until he eat of
the prey and drink the blood of the slain ! " he exclaimed, " What hath God wrought ! " so now, when
the prophecy has such a striking fulfilment, as our
forces, almost without rest, or food, or sleep, *sweep
on, aroused, resistless, dominant,* let us repeat, " What
hath God wrought! what hath God wrought ! "

Everybody admits, consciously or unconsciously,
that God's hand planted the nation.   He prepared
the seed; he prepared the soil.   He controlled the
forces of Nature, and the will, and the wickedness of
man, to the end that Plymouth Rock should be the
landing-place of our fathers, and Puritan principles
the foundation on which we should build ; and whether the people of this nation purpose and execute, or
purpose and fail, the building is going up.   God

knows how to make the wrath of man praise him.
The model of this nation has been determined by
the great Builder, and made certain by the principles which he has already wrought into it. Virgil,
the first of Roman poets, represents the goddess mother of Æneas presenting him with a celestial armor
with which he is to triumph over his enemies, and
establish the foundations of imperial Rome. He ex-
amined it all with intense interest ;

> " But most admires the *shield's* mysterious mould,
> And Roman triumphs rising on the gold ; "

for on that shield, celestial hands had prefigured the
destiny, the course and glory, of Rome.

When a tabernacle was to be built in which the
Israelites should worship, Moses was to have control
of the work, as indeed he had of every thing which
that nation did ; and the people were voluntarily to
supply the materials ; but it was to be built, *the
whole* and *every part* of it, according to a pattern
wrought in the Heavens. "After the pattern that was
showed thee in the mount, even so shall ye make it."

So for us, whether *we* see it or not, before celestial
eyes, wrought by the divine hand, our history is writ-
ten in prophecy,—and not the grand outline merely,
but the perfect model and all the working plans of
the glorious structure completed ; and just according
to that pattern all the resistless forces of this nation

are building. Who has not felt again and again during these last four years that we were putting the mighty efforts and sacrifices of the nation as materials for our future into unseen hands; that as the two great armies, alternately aggressive and receding, ebbed and flowed back and forth between the two capitals, an omnipotent hand held the scales, and weighed the issue of battles? Was God present with the people whose deeds the Old Testament records? Then he has been present with us. Was God present when Israel and Amalek fought, and the battle fluctuated in the wilderness and Plain of Rephidim,—present to give Joshua the victory? So has he been present with our armies. Did he have respect unto the rod of Moses as he lifted it high towards heaven in prayer?—and when his hands were weary, and his voice faint, and the enemy prevailed, did he acknowledge the renewed exertions and united supplications of Aaron and Hur? Then has he been with us, answering our prayers and succeeding our endeavors. Over all these battles, over all these humming messengers of death, over all these night assaults and defeats, over all these weary, exhausting marches through swamps and forests and mountain-gorges, God has presided, rifting results as huge stones from the quarry, and hewing and fitting them as materials for the great structure which is rising,—rising according to the pattern shown in *the mount*.

Let us thank him! let us pitch our songs to reach
his ear! let us praise Israel's God, our fathers' God,
and our God, for unexpected, thorough, complete,
and glorious victory! " The Lord is my strength
and song ; and he is become my salvation."

Let us thank God for the gift and endowments of
the Lieutenant-general! Oh! how easy it seems to
conquer now ; but let us call to mind the feeling
which blanched our checks, and bowed in the dust
our hearts, when, after long and confident expectation, in campaign after campaign, the only despatch
that stayed us up against immediate despair was the
poor mockery of our grief, " The Army of the Potomac is safe;" that is, safe in its retreat, safe in its in-
trenchments again, safe from immediate annihilation.
All unused to war as we were, soldiers could not be
made in a day, much less a commander. Besides,
any man who is physically sound—and most are till a
draft is ordered — may soon learn his duty and per-
form it. But who is to guide these forces when
armed and disciplined? Who is to employ, with plan
and skill and foresight, the mighty hosts scattered
over the country, and time their movements, and
turn their concentric exertions to account in securing
the one great end? Here is a demand for the highest order of mind, and for that peculiar combination
of intellectual and moral endowment which has ap-
peared only at rare intervals in the history of the

world. How long did England seek before she found her Wellington? France has had but one Napoleon. And who but Washington showed the necessary qualifications of a captain during all our long struggle for independence? The times do not make men, they call for them: and the more you think of it, the more deeply will you feel how absolutely dependent upon God we have been for such a man, and for getting him into the commander's position so soon. Of course, no man can predict his future, nor is it necessary. Enough that, bold in conception as he is reticent in speech, quiet in person as he is efficient in command, modest in appearance as he is unconquerable in purpose, he has been endowed, and raised up, and put in command, to serve and save the nation. Cyrus was not more the gift of God to the Jews than is Grant to this nation. U. S. G. means hereafter the *United-States General.* Let us appreciate God's goodness in furnishing us this leader, and be thankful, — thankful to the Creator and Preserver and Disposer of men and nations!

Let us thank God also for the brave men who have executed his plans and achieved the victory. Oh, what courage to go forth when certain death stared them in the face! Oh, what a spirit of self-sacrifice to close up broken ranks, and press steadily forward across ground once, twice, thrice lost, tramping in the blood of the dead and dying, and over friend and foe! Would that those titled and untitled heroes could have seen this day, and lived to enjoy the victories which their valor helped to win, and the new future which their blood helped to inaugurate! Such unfaltering courage from the Rapidan to the James, from Nashville to Atlanta, (a whole army dependent on one exposed rickety railroad, which was like a hundred men clad in marine armor going down into the deep sea to fight unknown forces, and only one little, brittle tube for all to breathe through,) and now from the Confederate capital to the capture of the Confederacy, — such dauntless courage, such almost superhuman endurance, such continued self-sacrifice, — let us thank God for such soldiers! And, remembering the patience, fidelity, and heroism of the navy, for such sailors too!

Mark also what a spirit God has given the people. How bravely they have sustained the Government! How patiently they have borne defeat and loss! How cheerfully they have anticipated the drafts! How readily they submit to taxation! How submissively they mourn their slain in battle! Neither Rome nor Sparta knew nobler fathers and mothers than weep to-day among our New-England hills and over our Western prairies. The God of all consolation comfort them!

Mark also how God has disposed the nations towards us, compelling us to develop our own re-

sources, and to depend upon own energies, keeping
us clear of all foreign entanglements and hostile in-
tervention.

But I must leave the story of God's goodness half
told: I can only hint at the multitude of mercies
which should call forth our praise, — our praise to
Him " who is glorious in holiness, fearful in praises,
doing wonders."

In conclusion, my friends, what shall we render
unto the Lord for all his benefits towards us? Shall
we not serve him as these soldiers have served their
country? This is what we want now, in the Church
as in the Republic, only bloodless, and, without carnal
weapons, a grand triumphant advance. You all re-
member the exhortation of the Apostle, " No man
that warreth entangleth himself with the affairs of
this life, that he may please Him who hath chosen
him to be a soldier." Were our self-devoted soldiers
under obligations to the country that we are not under to our God? If this war, on our part a holy one,
is only subservient to the far-reaching designs of Jesus, our captain and king, what shame ought to burn
the cheeks of many in the ranks of the Church! Entangled with the affairs of this life, some of us can
just make out to render once a week, a poor half-
day's service on the sabbath, manifestly having no
purpose or principle that does not govern an enemy
or alien to the cause of Christ. Entangled with the

affairs of this life, whole divisions are seldom present at the regular weekly appointments of the church, and never at an extra service, though conquests and captives wait their hands in our Great Captain's cause. Ah! if many of you who have enlisted in the ranks of the Church should bring yourselves to the standard of the New Testament, as you ought, or to the standard of that army discipline which makes conquering soldiers, one-half of you would condemn yourselves as stragglers and deserters, deserving no share in the successes of the Church, as manifestly, consulting your own ease and convenience, you bear no part in her sleepless vigilance and protracted conflict.

When would our armies have won without discipline, self-denial, and heroic self-sacrifice? Shame! shame on such indifference, laxity of principle, skulking, and cowardice as is found creeping into the ranks of the Church!

This world is in revolt against the Son of God. It is to be conquered and subdued to his righteous sceptre. Already the battle is joined. Unwavering devotion, prompt obedience, patient faith, and persistent purpose, — these are the elements that win. Let us each hear for himself the clarion call of the great warrior apostle ringing along our lines like the echoes of a bugle-note among the hills: "Thou, therefore, endure hardness as a good soldier of Jesus Christ."

Then shall the Church, like the Republic, advance
to new conquests and possessions; and our sabbath-
songs thrill like a battle-hymn on a nation's anniver-
sary. Then shall we sing, more sweet, more loud,
"The Lord is my strength and song, and he is become my salvation: he is my God, and I will prepare
him a habitation; my father's God, and I will exalt
him."

I Chronicles xxii. chapter, 6–16 verses. I will not transcribe the passage, but only so much of it as is necessary to exhibit its spirit and the foundation of my discourse.

"And David said to Solomon, My son, as for me, it was in my mind to build a house unto the name of the Lord my God; but the word of the Lord came to me, saying, Thou hast shed blood abundantly, and hast made great wars: thou shalt not build a house unto my name. . . . Behold, a son shall be born to thee, who shall be a man of rest: he shall build a house for my name.

"Now, my son, the Lord be with thee, and prosper thou and build, — only the Lord give thee wisdom and understanding, and give thee charge concerning Israel, that thou mayest keep the law of the Lord thy God. Then shalt thou prosper, if thou takest heed to fulfil the statutes and judgments which the Lord charged Moses with concerning Israel : be strong and of good courage ; dread not, nor be dismayed.

"Now, behold, in my trouble I have prepared for

the house of the Lord a hundred thousand talents of
gold, and a thousand thousand talents of silver; and
of brass and iron without weight; timber also and
stone. Moreover, there are workmen with thee in
abundance, — hewers and workers of stone and timber, and all manner of cunning men for every manner of work.

"Of the gold, the silver, and the brass, and the iron,
there is no number. Timber and stone also, all is in
abundance. Arise therefore, and be doing, and the
Lord be with thee."

Like David, our period of war and bloodshed is passing away. Like Solomon, a new period of peace and
rest, its offspring and successor, is about to succeed.
Generally, it is true that every era is the child of
that which preceded it; but the era upon which this
nation is now entering is peculiarly and strikingly
the offspring of that which is passing away. The
present, these four years and more of pain and agony
and travail are not merely giving place to the peace-
ful future, — this closing present is the parent of the
hopeful, matchless future that is just stepping into
its place.

And upon this rising future is devolved the inspir-
ing but tremendously responsible task of building a
national structure whose magnificence and glory
shall be to that which preceded the war, as the temple of Solomon was to the tent or tabernacle in which

David worshipped. Using the text analogically, the points of discourse to which your attention is invited are two; viz., the materials accumulated, and the laws which must govern a successful use of them.

1. And first, as to the materials secured and prepared for future use by this period of war. After the boundaries of a nation are settled (and we are now pretty well defined in this particular, I think; at any rate, between the oceans on the *east* and *west*, and between the lakes on the north and the Rio Grande on the south, we are not likely to have any hostile or divisive lines run for some time to come), — after the boundaries of a nation are determined, the next question relates to the materials, — the materials of self-development, of growth, and greatness; and these, if the nation is to attain prominence and permanence, must be abundant and varied. Not for the building of the temple alone, but for the building of the nation as well, there must be gold and silver and brass and iron and stone; and was ever a nation more abundantly provided than we are with all rich mines of precious metals, minerals, and ores? As to our gold mines on the Pacific coast, we have been in the habit of congratulating ourselves for the last fifteen years that they were kept concealed till those western slopes, which connect the lofty line of perpetual snow with the warm waters of the gentle

Pacific, came into the possession of a Protestant Christian nation.

But does it not seem as if the value of those mines was reserved especially for these days of war and bloodshed? What should we have done without them in the disturbed state of our currency? Where should we have obtained the vast amounts of gold that we have been obliged to export while we were issuing paper and promises to pay at such a fearful rate? What a timely deposit these mines have proved!

And, so far from being exhausted, it would seem that we have but just begun to comprehend their extent and value. We have gathered up what has been washed out upon the surface. The great vaults hidden in the mountains, and the richly veined avenues that lead to them, are yet to be opened. During these years of bloody war, we have learned much as to the value and position of these deposits, and, therefore, are prepared to draw against them for the future.

Great advances have been made, too, in the discovery of silver, of copper, and lead and iron.

Of petroleum what shall be said? Who could have divined, four years ago, that seas of oil should be discovered in the interior of the earth during this tremendous civil war? — an oil of so many and such remarkable qualities, that we have but just begun to conjecture the uses to which it may be put in the economy of civilization.

Doubtless, many a property and many a company are alike bogus; and yet, after due allowance for all speculation and gambling, the real discoveries of precious metals, minerals, and ores, have been many and of immense value; and around all these mines must be gathered from overcrowded centres and cities a population, more or less dense, of capitalists and laborers. Here, at the base of the mountain, in the forest never traversed before, except by the Indian and the buffalo, manufactories will rise, villages spring up, and fields be cleared; and whoever secures the building of a house, or a barn, or workshop or factory, or the converting of a tract of wild land into a farm on the coast of the Pacific, or in the forests of Pennsylvania, adds to the wealth and prosperity of the nation.

Our agricultural resources were large and rich before; but we add to them immensely by the issue of this bloody war. Now the sunny South is open to free labor and all the healthful rivalries of personal productive industry. Here are vast plantations to be made as the garden of the world under the economy of the Northern farmer; here are unmeasured wastes of virgin soil waving in an unshorn wantonness of tropical luxuriance; and broad meadows white with cotton from generation to generation;

and immense swamps reeking with a heat and mois-
ture that may grow rice for the world. All these
plantations, wastes, wildernesses, meadows, swamps,
and unexplored lands stretching from the Chesapeake
Bay far down to the Gulf of Mexico, and around
it, are to be divided, like our Northern lands and
Western prairies, into innumerable farms and free-
holds occupied and improved by the willing, intelligent owner, white or black, and thus to become
an incalculable means of growth and wealth to the
nation.

As regards all such materials of national greatness,
the war closes with nothing wanting. Leaving now
the productiveness and extent of the soil, and all the
minerals, mines, ores, metals, deposited beneath its
surface, let us turn to the people dwelling here. This
is a material without which the wealth of the mines
and soils are of little account. Manifestly, there is
a great difference in peoples; one nation "living out
feebly and obscurely the term of its existence, and
then sinking into complete oblivion; another exer-
cising a powerful influence over its contemporaries,
and leaving a luminous track in the annals of the
world." Lord Bacon, speaking of the true greatness
of kingdoms, says, "The greatness of an estate in
bulk and territory doth fall under measure; and the
greatness of finances and revenue doth fall under
computation. The population may appear by mus-

ters, and the number and greatness of cities and towns by cards and maps; but yet there is not any thing amongst civil affairs more subject to error than the right valuation and true judgment concerning the power and forces of a State."

"Walled towns, stored arsenals and armories, goodly races of horse, chariots of war, ... ordnance, artillery and the like,—all this is but a sheep in a lion's skin, *except the breed and disposition of the people be stout and warlike.*" Just at this point was our weakness. Our pursuits, our education, our love of wealth and ease, had unfitted all the inhabitants of the free States for the perilous adventure and bloody carnage of war. We had every thing else, but we had not the *stout disposition* and *warlike temper.* We had become broad of forehead, but narrow of chest, intellectual and far-sighted, but weak of nerve and faint of heart. The dollar, so grateful to the palm, had unfitted it to clasp the sword and the musket. Our young men sought places behind the counter, avoided the out-of-door toil, the frown of the elements, and the heat of the sun; and so they had come to have neither strength for the camp nor courage for the fight. They could not and they would not go to war. So multitudes believed; and thus our Government was like a rich prize, waiting only a mailed hand to pluck it,— such was the talk on the floor of Congress, such was the expectation of our enemies;

but I need not say how completely the war has revolutionized and corrected all this. The spirit of '76 had not died with our fathers. The heroic spirit of Spartan mothers was in our women; and the endurance of Roman soldiers in the men that answered to the country's call.

We have discovered or developed that stock and spirit, and marvellous power of adaptation, which, with the blessing of God, insures the perpetuity of our Government and our free institutions against all secession and revolt at home, against all interference and insult from abroad.

All unprepared for such a gigantic and formidable rebellion, and abhorring the carnival of war, we have, nevertheless, created a navy which exacts the most deferential regard of the nations; and, turning the country into a camp, and the makers of cars and pianos into the makers of cannons and rifles, we have tried and settled the question of self-preservation; and it will be a new element in our future,— a truth demonstrated and priceless, that the Republic wants neither living patriotism, nor the people warlike power.

And, in settling the question of self-preservation, we have also settled the question of capacity,—the capacity of the people to govern themselves. However clearly the *right* of self-government may be deduced from abstract principles, to maintain that right

successfully, without the *capacity* of self-government,
is clearly impossible and absurd. We had kept along
well enough in peace, as the most unseaworthy craft
may keep her keel in smooth water, and the most in-
competent pilot hug the land in fair weather. Up to
1861, it was an open question, whether the people of
this Republic were capable of self-government. Then
came the storm and test ; and who was certain of the
issue? Who had faith in the people then? Some
voices will be very loud *now*, telling of courage and
confidence and prophecy; but no man heard them
in 1861. We listened in vain for a clear, command-
ing utterance from any quarter. No man knew what
to do. Those who should have directed and led
the way stood paralyzed. Senators left their seats.
States seceded and organized within the limits and
jurisdiction of the United States. Our statesmen
faltered, — *made defenceless*, apparently, by the Constitution which they had sworn to *defend*. Our politicians *babbled* more than usual. The Old World looked
from her thrones across the water, and said, " Yes,
yes, I told you so : the great bubble has burst ; your
Republic has gone up !"

But by and by Beauregard opened fire on Sumter,
humbling the flag, which, after four years, in the shot
and smoke of battle, goes up again to-morrow, to float
in triumph over the old fort, and over the humiliation
of every traitor that has ever lifted his hand to assail

it ; *and that fire was something that the people could understand and decide about :* that was an act of war which went booming through the land, waking a power that instantly took control of the Government,—comprehending the issue, and rushing volunteers into the ranks by the hundred thousand. Resistance — instantaneous, unconcerted, sympathetic, inflexible resistance — evinced the unity, the instinct, and the purpose of the people ; and in all these four years, tried sorely, disappointed in the repulse and defeat of their armies, impatient of delays, the people have never faltered. Politicians faltered ; many of the chief rulers began to think of retreat, of some kind of settlement ; newspaper editors cried, Enough!—enough of blood, enough of drafts, enough of debt!—and hied them away from the great city to the Canada shore, by the Falls, to plot something, to do something, they knew not what ; but the people, following their instincts, held steadily on their way as if inspired by the Almighty. There was no wavering, no signs of weakness, no want of capacity, with the people.

In addition to all this, and adding immensely to the strain, in the very midst of the storm and fury of civil war, a Presidential election came. The polls were opened ; bayonets did not guard nor govern them. The day passed as quietly as the sun runs his course, and closed with benedictions and doxolo-

gies; *and the capacity* of the people for self-government is proved. The question of capacity is settled
and recorded forever. A gifted and worthy son of
Massachusetts, speaking at a time when every
mind was burdened with apprehension, says, "If the
people shall go through this election freely and fairly,
whatever may be the result as regards parties, our
institutions will have received a triumph: no nobler
spectacle will have been witnessed in this land, since
it first asserted its title to be called a land of liberty."
Behold, the triumph, the nobler spectacle!

Another question, and one of vast importance in
this nation, has been settled; and the conclusion, defi-
nite and dominant, is one of the grandest materials
to be built into our rising future: I mean the question of *State sovereignty.* Out of this doctrine of
State sovereignty comes the claim to declare null
and void a law of the United States, — out of this
comes secession and war. This suicidal doctrine of
State sovereignty is not in the Declaration of Independence: that reads, "We, therefore, the represen-
tatives of the United States of America, . . . *by the
authority of the* GOOD PEOPLE *of these colonies,* solemn-
ly publish and declare that these United Colonies are,
and of right ought to be, free and independent
States." Mark two things: 1. This is done by authority, not of the States, but of *the people.* 2. The

States are declared " free, independent," but not sovereign.

The doctrine was in the old confederation, which lasted thirteen years from the time of the Declaration of Independence, and was a failure.

The doctrine is not in the Constitution. That starts off, like the full sweep and swell of the sea : " We, *the people* of the United States, . . . do ordain and establish this Constitution." Our national Government is not the creation of sovereign States, — is not a compact, or league, to which they may accede, and from which they may secede. This is the " people's Constitution, the people's Government, made for the people, made by the people, and answerable to the people ; " and this same people, and not sovereign States, have declared this national Government supreme, and the Constitution the supreme law of the land. " This Constitution . . . shall be the supreme law of the land, . . . any thing in the constitution or laws of any State to the contrary notwithstanding. John Quincy Adams says tersely, speaking of this doctrine of State sovereignty, " The Declaration says, It is not in me. The Constitution says, It is not in me."

And yet, this very doctrine, having no fixed foundation in our history, and none in the great exponents of our national life, and absolutely destructive of the very idea of self-preservation, exposed and exploded

by Mr. Webster in his reply to Hayne, and again
triumphantly denied as false, and denounced as suicidal, in his reply to Calhoun,—this doctrine, like the
genius of evil, has ever been intruding its bastard
presence through all our history. No argument could
expel it from the minds of Southern men. They
wanted to believe it, and they would believe it. From
history, from argument, on the floor of the senate,
from the judgment of the people, they appealed to
the bloody arbitrament of the sword; and the decision
is again, as always, against them,—triumphantly,
overwhelmingly against them. There is no other
appeal. Beaten on every field; compelled to surren-
der their armies as prisoners of war; in the midst of
all the poverty, mourning, anguish, and desolation,
which they have brought upon themselves, compelled
to see the old flag raised again upon the forts, arsenals, post-offices, and custom-houses of the United
States,—they are not likely to repeat their appeal
to the sword. The question of State sovereignty is,
therefore, settled: and the settlement is a corner-
stone in our new structure.

And the settlement of this question settles the
policy of the Government and the destiny of the
South. The Government must sustain the loyal
men of the South, whatever their social condition or
color. Any other hypothesis is too base to be enter-
tained. Multitudes of these rebel leaders have fled,

abandoning their homes, their social position, their
property. Loyal men, having no occasion to fly, have
remained, and must occupy these vacant places and
this abandoned property. Already this transfer has
been made largely, is now being made rapidly, will
continue to be made heavily; and these loyal
men, established directly by the Government on these
rice lands and cotton plantations, or in the cities
through the failure of rebels to pay their taxes or re-
deem their estates, — these loyal men, having come
into the possession of the forfeited social position and
property of the guilty, must receive three things; viz.,
1. The jealous protection of the Government. 2. The
unqualified support of the Government. 3. And the
immense patronage of the Government. And then,
further, if human nature can be relied upon as a basis
of reasoning, these men will appeal to each other —
those who are there, and those who shall go there —
for mutual support; and they will appeal to the Government against the restoration, and against the return, of those known by their neighbors to be guilty,
as well as against those who have confessed their guilt
by flight, or by the arms captured in their hands.
Let us not be disturbed, therefore; for there are causes
at work, whatever may be the leniency of our national authority, — causes that have gone too far to be
resisted or controlled. — causes which must decide
the future of the South. Let Congress keep strictly

within its power to inflict punishment for treason,
and maintain that "No attainder of treason shall
work corruption of blood, or forfeiture, except for
life." Those leaders in rebellion are to be leaders no
more.   Passing those who will pay the penalty of
their crime with their lives,—as some of them surely
must, or all crime go unpunished,—not one of those
men who sat in the councils of the nation, nor one of
those who, with perjured hands, drew their swords
against the Government that had educated and sup-
ported them,—not one of those traitors, nor their
children, will ever regain their former position.   The
power is neither with *them* nor *their* constituents.
It has passed at the fountain into other hands.   The
road to their old supremacy is held, at the popular
end of it, by other people; and the very gate of en-
trance controlled by new views, new doctrines, and
new sympathies.   These Southern men have yet to
learn, and possibly a few others also, how completely
they have revolutionized their own society.   Hence-
forth, the Government is a unit, controlled and worked
in the interests of loyalty, freedom, and nationality;
and this will be found another and a mighty element
in our rising future.

Such, my friends, as well as I can describe them
now, are some of the elements or materials gathered
and prepared by this war for the building which is
to rise in the days of peace upon which we are en-

tering; and they are as essential and as abundant certainly as the materials prepared by David for the building of the temple.

I have been compelled to write in weariness and in haste; but I cannot conclude without touching an essentially practical thought for us all. How shall these materials be put together? on what foundation? in accordance with what laws? beautified with what ornament?

1. First as to the foundation. The temple, which is the glory of Jerusalem, rested on deep foundations of solid rock; and so our vast national structure must rest upon an immutable foundation; that is, upon the fundamental doctrines and principles of the inspired *word*. The Christian religion is the only secure basis on which we can build.

Expediency has its place in the affairs of Government; but, as a foundation on which to raise a vast and permanent superstructure, it is a quicksand. Neither can morality alone bear the tremendous weight of Government. Religion—the religion of the New Testament, vital, immutable, eternal—is our only foundation.

Washington, in his Farewell Address, says, " Religion and morality are indispensable supports of every habit and disposition that can contribute to national prosperity;" and then adds like a discriminating preacher of the gospel, " and let us with caution indulge the supposition, that morality can be maintained without religion,—morality cannot prevail in exclusion of religious principle." And then he adds with a power of exhortation which has lost nothing of its point or pertinency, " Who, that is a sincere friend to the *Government*, can look with indifference upon attempts to shake the foundation of the fabric?"

The declaration of the prophet has been mournfully illustrated in the downfall of many and mighty structures: " For the nation and kingdom that will not serve thee shall perish." Jesus Christ is the corner-stone of our national as of our personal salvation.

2. Then again: in accordance with what laws shall we build? There is no objection to a leaning tower, as an object of curiosity to attract the traveller; but the walls that are to stand must rise in accordance with the laws of gravitation: not more surely must the fabric of the nation rise in accordance with the laws of God. " This way of tying walls together with iron, instead of making them of that substance and form, that they shall naturally poise themselves upon their butment, is against the rules of good architecture," says one; and George Herbert has this line: " Houses are built by rule, *and commonwealths.*"

To determine what the will of God is, as respects society, and to define that will in laws to which the judgment and conscience of the people will respond, is the great work of the legislator. To re-affirm the ordinances of Nature, to re-enact the laws of the Creator and Sovereign and Judge of the world,— this is the essence of all wise legislation.

We are indeed through the war; but we have a mighty work on our hands,—one over which, in the midst of all our joy, we may well fast and pray,—a work requiring the profoundest wisdom, the firmest adherence to principle,—a work that can only be done in obedience to the commands of God.

This is what we must do, — all the people of this land, — render the hearty, scrupulous obedience of Christian men in all our work. No mistake is more fatal than that a great nation, with such materials as we have, can afford to be lax and lawless and irreligious. Just the reverse is true. Any thing that will live, endure, must render an obedience stricter as it is greater. For instance, a grain of dust may not always and uniformly obey the laws of gravitation; but the heavenly bodies, sun, moon, and stars, must obey. The great ocean must obey influences that little lakes and rivers are not required to recognize even.

Just so, like the sun and the sea, this nation must obey the laws and precepts of the Almighty. Let us each one, with heart and hand, with lip and life, however humble, however lofty our part, do all in strictest obedience to Him who filled the temple with his glory, and honors them that honor him!

3. And, finally, as to the adorning of our temple: above all the splendor of art, above all the creations of genius in stone or on canvas, let it be like the inscription on the bells of the horses in the last days;—on all our door-posts and windows, on ceiling and walls and floor; let it be, "The beauty of holiness;" "Holiness to the Lord;" and let it be reflected by all the builders, "The ornament of a meek and quiet spirit, which is, in the sight of God, of great price." "Thus saith the Lord of hosts, . . . the fast of the fourth month shall be, to the house of Judah, joy and gladness and cheerful feasts: . . . therefore love the truth and peace."

And do thou, O Lord, for whose sake we would build our temple enduring, lofty, and beautiful, "make us glad according to the days wherein thou hast afflicted us, and the years wherein we have seen evil. Let thy work appear unto thy servants, and thy glory unto their children. And let the beauty of the Lord our God be upon us: and establish thou the work of our hands upon us; yea, the work of our hands establish thou it."

THE ASSASSINATION OF THE PRESIDENT.

He calleth to me out of Seir, Watchman, what of the night? Watchman, what of the night?

The Watchman said, The morning cometh, and also the night.

THESE words seem to me strikingly appropriate to our present circumstances. Last Sabbath morning it was my privilege to place before your minds some reasons for thankfulness, — thankfulness to God. Then the streets were decked with symbols of joy; gladness in welcome accents broke from every lip. Men's countenances were bright, as if reflecting the coming of the morning. We clasped each other's hands with a jubilant pulse, and every eye answered back hope, inspiration, to the eye that looked into it.

But how changed is all in a moment! Yesterday morning flags were set at half-mast. Even Sumter's flag is but half raised. As the day advanced, emblems of mourning drooped from the highest windows to the sidewalk. *The President is assassinated!* Men hold their breath, and turn pale at the appalling words. Citizens meet, and

shake hands, and part in silence. Words convey
nothing when uttered. All attempt to express the
nation's grief is utterly commonplace and insignifi-
cant. An eclipse seems to have come upon the
brilliancy of the flag, — a smile seems irrelevant
and sacrilegious. Even the fresh, green grass, just
coming forth to meet the return of spring and the
singing of birds, seems to wear the shadows of twi-
light at noonday. The sun is less bright than before; and the very atmosphere holds in it, for the
tearful eye, a strange ethereal element of gloom.
Surely " *the night cometh.*" And as we gather here
this morning, after an absence of only two days,
how appalling, in this cheerful home of our religious
affections, are these wide-hung emblems of grief and
anguish! It is manly to weep to-day. The coming
of the morning, and also the night, are strangely
mingled.

Had death overtaken any one of our brilliant
military leaders in the field, we should have said
it was a thing to be expected. Had any sudden
reverse in the fortunes of war visited one of our
armies, it would have been a terrible grief, but still
a kind of calamity to which we have become accus-
tomed. Had the President fallen by a chance
shot in Richmond, or by the hand of some lurk-
ing assassin, as he passed the fortifications through
which our hearts did not consent to his going,

we should but have realized some of our transient forebodings. But after his safe return. and the triumph of our arms, which he took so much pleasure in telegraphing to the people, we had almost dismissed from our minds any fears for the safety of his life. And hence the telegram announcing the death of the President at such a time, in such a way. falls upon us like a crash of thunder from an unclouded sky.

Wearied with the duties of his high position, and the persistent annoyance of petty office-seekers. and unwilling to disappoint the people even in their unreasonable expectations. he sought an hour's recreation in the theatre. And what a horrible tragedy! The actor, having thoroughly prepared his part, and being often defeated in one way and another from the fiendish acting of it, finds his opportunity at last. With the stealthy step of a base, brutal coward, with a damning lie on his tongue, and the heart of a demon in his breast. he approaches the generous. unsuspecting man in the rear of his seat, and. aiming the fatal weapon with practised hand at the back of his head. puts the ball directly through his brain. and then makes his escape through the screens and drapery and doors with which his calling had made him acquainted. There are no last words for wife or children. — no word for the people's heart to which he always spoke. — no parting counsel for a bereaved and almost bewildered nation. The hand that signed the *emancipation proclamation* hangs helpless in death; the mind which had borne so evenly the tremendous strain of four unparelleled years is hurled from its throne; the great, good, magnanimous heart is stilled; those generous lips which have spoken *mercy* so often, and would perhaps, like martyred Stephen's, have said in their last articulate speech, "Lord, lay not this sin to their charge," are sealed forever. The nation has lost a father; the human race a sincere, devoted, and able leader!

I have had no time to analyze the character, or choose out words to express our sense of the worth, of the late Abraham Lincoln. But I may employ, with your approbation I am sure, the words used by Daniel Webster concerning Zachary Taylor: "He has left on the minds of the country a strong impression; first, of his absolute honesty and integrity of character; next, of his sound, practical good sense; and, lastly, of the mildness, kindness, and friendliness of his temper towards all his countrymen."

Yes: "towards all his countrymen." He was, on the very day of his untimely death, exerting all the kindness of his unselfish nature, and prepared, it is believed. to peril all his great popularity. in inaugurating a policy *most lenient, most forgiving.*

towards those who had forfeited every thing except
the right to be hung. They have put aside their
friend. They have murdered the new-born mercy
which waited to bless them. No man could if he
would, and no man was disposed to, do so much
for them as Abraham Lincoln.

And how the loyal people confided in him! how
implicitly the common people trusted him! The
world has scarcely seen the like. He came to the
chair of the Chief Magistrate from the rough expe-
rience of frontier life. He owed his election, and
the favor with which he was received, to the be-
lief, in the minds of the people, that he was an
honest man.

And did he disappoint that confidence? Did he
show himself unworthy? Did he ever incur the
suspicion of dishonesty or corruption? Or did he
ever swerve from what he conceived to be the
path of duty to win popular applause? Never. On
the other hand, so impartial was he in selecting
men from all parties to fill the high offices of
government, so artless was he in all that he did,
so transparent were his deeds and his motives,
that, by a popular vote scarcely paralleled, the
people called him a second time to guide the nation for another four years. He knew nothing of
tricks or double-dealing or party-shifts, or crooked
policies. He was a sincere, impartial, straightfor-

ward, honest man. And the people saw it and
felt it, and were glad of an opportunity to honor
him with an overwhelming repetition of their well-
placed confidence. What a noble example is he to
all young men looking to office or popular regard!
With no military reputation, with no brilliant ora-
tory, with no winning grace of manners, he was
the foremost man for the highest office in the gift
of a great, free, and intelligent people, once and
again, because he was a man of absolute honesty
and integrity of character.

And besides these unselfish, impartial, upright elements of character, there was a masterful common
sense, a genial mother-wit, and a practical statesman-
ship, which showed themselves in some of the most
compact specimens of argument, happy avoidances
of difficulty, and a thorough apprehension of popular
instincts and judgments.

He was unpolished in style, but he was profound
in thought. He was pithy in his sentences, but original and patient in investigation; rough on the exte-
rior, but a jewel within,—

" Rich in saving common sense."

How much we owe to his unambitious example; how
much to his far-reaching discernment; how much to
his good-natured hearing of all sides; how much to
his steady calm judgment, which held the scales, in

the fury and gusts of the storm, as equally poised as
if in the atmosphere of peace and calm; how much
to his great forbearance under stinging reproach;
how much to his knowledge of and unwavering confidence in the people and the people's cause, God
knows, but we know not as yet. May the day never
come, when, by bitter contrast, we shall learn how
wise and safe was the confidence which we reposed
in him!

This nation mourns to-day as it never mourned before. The statesmen of the land had learned to trust
him in the greatest exigencies; the impatient were
restrained by his moderation; the immovable and
morose were moved and almost brought into time by
his steady, sympathetic step forward; the one-eyed
were made ashamed of their ignorance by an hour
in his society; the revengeful learned magnanimity
from his deeds. The soldiers loved him, and the soldier's mother loved him, and confided in him. The
negroes loved him; oh, how they will mourn for him!
Moses was not allowed to lead the children of Israel
into the land of peace and plenty; neither was he
allowed himself to enter it, but only to survey its
broad prospect from Pisgah's top. And so *their* de-
liverer and *ours* is only permitted to come to the
border, and, in these last few days, catch pleasing
glimpses of the glorious, opening future. And as
when Moses died, his eye not dim, and his natural

force not abated, there was mourning throughout all the camp, till the plain of Moab resounded with the cry of sires and sons, mothers and maidens; so now there will be mourning in the camp, and mourning on the prairies, and far away over the mountains; but nowhere keener anguish and disappointment than among the sable hosts whom his noble heart and hand has freed. All men unconsciously speak of him as our beloved President; and the hand of the assassin has embalmed him with all his virtues and greatness, and made him sacred and sublime in our fond, loving hearts, and in the heart of the world forever.

Were I to select some one thing by which to characterize Abraham Lincoln, I should name his profound apprehension and appreciation of the popular instinct, —that instinct which is true to the right as the needle to the pole, in all storms, and on every sea. He believed in God. He believed God was to be recognized in this war. He believed that the *set* of the loyal masses,—the deep, silent current, which bears on events, is in the line of God's advance; and, thus believing, he governed himself by his apprehension of the people, and of God, as manifest in their silent set or drift. As the philosopher learns the plans of God from an unprejudiced study of Nature, so he learned the purposes of God from the instincts of the people. As the naturalist discovers from the structure of the

animal what its mode of life and habits must be, so he
saw from the essential peculiarities of our Government whither our future must tend. He did not
mean to be ahead of the popular feeling; for then
there would be a re-action against his policy. He
did not mean to be much behind it; for then some
other agent might be sought, through which to give
it expression : and so, regarding the voice of the loyal
people, in this great crisis of the Republic, as the
voice of God, he kept his ear open and his eye attent,
and marshalled his policy not quite abreast of the di-
vinely led masses. He sought not to control an age
thus moved and inspired, but to be controlled by it.

Herein was his wisdom, herein his greatness,
herein his power. This was the secret of his success,
the source of that light which, in all coming time,
shall gild with unfading splendor the name of Abraham Lincoln.

As the Netherlands mourned for William, Prince
of Orange, as France mourned for Henry IV., "We
have lost our father, we have lost our father!" so
America mourns to-day.

> "Such was he; his work is done;
>   But, while the races of mankind endure,
>   Let his great example stand
>   Colossal, seen of every land;
>   And keep the soldier firm, the statesman pure,
>   Till in all lands and through all human story,
>   The path of duty be the way to glory.

But speak no more of his renown,
Lay your earthly fancies down ;
And in the vast cathedral leave him,
God accept him, Christ receive him!"

1. And now, my friends, what are the lessons of
this great calamity ? *First of all, submission.* God
reigns : we are absolutely dependent and sinful. The
Emperor Mauritius seeing all his children slain before
his face at the command of the bloody tyrant and
usurper, Phocas, himself expecting the next stroke,
exclaimed aloud, in the words of David : " Righteous
art thou, O Lord, and upright are thy judgments."
This event takes us by surprise ; but the origin, ma-
turity, and perpetration of this awful crime, was all
under the sleepless eye of God. For reasons which
we cannot fathom now, nor find, he has permitted
it. Perhaps, when this day, *the 14th of April*, forever marked in our calendar ; marked by the humbling of the flag at Sumter ; marked by the exaltation
of the flag four years after, — perhaps, when the
14th of April comes round four years hence, we shall
know more of God's designs in permitting this foul
murder of our beloved President. There is ONE
whom the hand of violence cannot reach ; and he
has not led us thus far to desert and destroy us now.
Meanwhile, as becomes us, let us bow our heads in
meek submission to the divine will! Surely his foot-
steps are in the great deep ; his designs are hidden

from us in the dark; but let us trust him; let us cleave unto him! Submitting penitently to the rod of affliction, let us put our hand in his, and say, Father, lead; Father, spare and bless!

2. A second lesson is this: *Execute justice in the land.* What is the foundation of our confidence in God? Is it not that he will do right? Is it not what David says over and over again in all his trials? —*justice* and *judgment* are the habitation of his throne. And just these—*justice* and *judgment,* are the foundation of every throne, and of every government. I spoke on Thursday. as far as it was appropriate to my theme. of the tremendous mistake and folly and sin for the people of a great nation to think that they can neglect or violate the laws of God with impunity. Just here has been our danger. There has been a miserable, morbid. bastard philanthropy, which, if it did not make the murderer's couch a bed of flowers, and set his table with butter and honey. made him an object of sympathy. and, after a while. of executive clemency. We are weak in our sense of justice. Why, how long is it since a man was pursued in the streets of Washington, and, though begging for his life, shot to mutilation? He was guilty of a foul crime? Yes; but did that give the injured man a right to murder him? Are there no courts, no ministers of justice, in the land? But the murderer was acquitted with applause in the

court-room. Only this very spring, a young woman shot one of the clerks dead in the hall of the treasury-building. To be sure, she said he had broken his vow to marry her! And, when I was in Washington a few weeks since, it was confidently expected that she, too, would be acquitted. And here in Massachusetts, not to speak of other States now, where the punishment of murder is death, the guilty wretch — who could brood over his infernal plan for weeks, and finally, after several attempts on the same day, execute it upon an innocent, unsuspecting young man, and all for the sake of a few hundreds, or, at the most, thousands, of dollars — is allowed to live, and become an object of sympathy. To shield his forfeited life imperils that of every young man who stands behind a counter in Massachusetts. Living, he is an encouragement to all persons like-minded to do likewise. Yea, saith the governor, ye shall not surely die.

And so in regard to the leaders of this infernal Rebellion: the feeling was gaining ground here to let them off really without penalty. They are our brethren, it is said. Then, we reply, they have added *fratricide* to the enormity of their other crimes, and are unspeakably the more guilty!

The punishment which a nation inflicts on crime is the nation's estimate of the evil and guilt of that crime. Let these men go, and we have said practically that treason is merely a difference of political opinion.

I do not criticise the parole which was granted; though, for the life of me, I cannot see one shadow of reason for expecting it will be kept by men who have broken their most solemn and deliberate oath to the same Government. It was not kept by the rebels who took it at Vicksburg. Nor will I criticise, for I cannot understand, the policy which allows General Lee to commend his captured army for "devotion to country," and "duty faithfully performed." But I considered the manner in which the parole was indorsed and interpreted as practically insuring a pardon; and to pardon them is a violation of my instincts, as it is of the laws of the land, and of the laws of God. I believe in the exercise of magnanimity; but mercy to those leaders is eternal cruelty to this nation,—is an unmitigated, unmeasured curse to unborn generations! It is a wrong against which every fallen soldier in his grave, from Pennsylvania to Texas, utters an indignant and unsilenced rebuke. Because of this mawkish leniency four years ago, Treason stalked in the streets, and boasted defiance in the halls of the capitol; Secession organized unmolested, and captured our neglected forts and starving garrisons. Because of a drivelling, morbid, perverted sense of justice, the enemy of the Government has been permitted to go at large, under the shadow of the capitol, all through this war. God only knows how much we have suffered for the lack of justice. And now to restore these leaders seems like moral insanity. Better than this, give us back the stern, inflexible indignation of the old Puritan, and the *lex talionis* of the Hebrew Lawgiver. Our consciences are debauched, our instincts confounded, our laws set aside, by this indorsement of a blind, passionate philanthropy.

Theodore Parker has a passage in his work on religion, in which he gathers into heaven the debauchee, the swarthy Indian, the imbruted Calmuck, and the grim-faced savage, with his hands still red and reeking with the blood of his slaughtered human victims. And the idea, to me, of placing the leaders of this diabolical Rebellion in a position where they might come again red-handed into the councils of the nation, is equally revolting and sacrilegious. It makes me shudder; and yet I think there was an *indecent* leniency beginning to manifest itself towards them, which would have allowed to these men, by and by, votes and honors and lionizing. The soldiers did not relish this prospect. They are not to be deceived by the misapplication of the term "magnanimity" to an act that turns loose into the bosom of society the men who systematically murdered our prisoners by starvation, and again and again shot prisoners of war after they had surrendered, — shot gallant officers, even

in these last battles, after being told that they were
mortally wounded, and strung up Union men in
North Carolina, because they had enlisted in the Fed-
eral army.

And now, *we* see and feel just as the soldiers do.
The spirit that shot down our men on the way to the
capital; the spirit that shot Ellsworth at Alexandria;
the spirit that organized treachery, treason, and rebellion; the spirit that armed those leaders to strike
at the life of the Government, — is the same hell-born
spirit that dastardly takes the life of our beloved
President, is the same atrocious spirit that seeks
the bed-chamber of a sick and helpless man, and cuts
his throat, and strikes the murderous dirk at the
heart of every attendant. We see its malignant,
fiendish nature now!

And what shall be done with these secessionists if
we succeed in arresting them before they get out of
the country, with the blood of the President, and of
the Minister of State on their hands? Pity them as
insane? parole them as prisoners of war? Doubtless,
like the St. Albans raiders, they have their commis-
sion from Richmond! Does this make your blood
boil? is this too shocking to suppose? Well, shall
we hang them? — hang the less guilty, and let the
more guilty go free? hang the miserable, worthless
hirelings, and let the principals and chiefs live? To
do that is to arm men, and goad them to take ven-

geance into their own hands. The instinctive justice of the human conscience must be satisfied by the action of Government, or it will have private revenge. There is a consciousness of right in the masses, that will not be tampered with in such a time as this. Not the branches of this accursed tree, but the trunk and the roots, must be *exterminated* from the land. Hear me, patriots, sires of murdered sons, weeping wives and orphans, — I say *exterminated!* " Ye shall take no satisfaction for the life of a murderer, and ye shall take no satisfaction for the life of him that is fled, that he come again to dwell in the land; for blood it defileth the land, and the land cannot be cleansed but by the blood of him that shed it." And, when David died, he charged Solomon to fulfil this divine command in regard to Joab and Shimei, who had been too strong for him during his life.

3. One thing more : *let us face the future, and all the solemn responsibilities of these uncertain hours, with courage.* We have God on the throne that no violence can reach, — the God who has always been with us. " Why art thou cast down, O my soul? and why art thou disquieted in me? Hope thou in God; for I shall yet praise him who is the health of my countenance and my God."

And then, such is the happy structure of our Government, that no assassination can arrest its wheels.

A terrible calamity has overtaken us; but it will only the more exhibit the inherent vitality of our institutions, and the greater strength of the people.

Andrew Johnson, who now becomes the Chief Magistrate by the mysterious providence of God, is unquestionably an able man. He has been much in public life, and never failed, except in his speech on inauguration-day, to meet the exigencies of his position. Besides, he has had a schooling in Tennessee which may have prepared him to lead at this very time. When I was in Washington, four years ago, I heard much in his praise. He told the secessionists, who were just then leaving their seats in the Senate to inaugurate the Rebellion,— told them to their faces, for substance: " Were I President of the United States, I would arrest you as traitors, and try you as traitors, and convict you as traitors, and hang you as traitors." And, judging from the speech which he made at Washington after the news of the fall of Richmond, he has not changed his mind.

We want no revenge: we will wait the forms and processes of law. We want justice tempered with mercy. We want the leaders punished, but the masses of the people pardoned. Let us confide in him as our President. And do you make crime odious, disfranchise every man who has held office in the rebel government, and every commissioned officer in the rebel army: make the halter certain to the intelligent and influential, who are guilty of perjury and treason, and so make yourself a terror to him that doeth evil, and a praise to him that doeth good,—and we will stand by you, Andrew Johnson.

Another ground of courage is, that the nation is a unit against rebellion to-day as it never was before. It is too much to hope,.I suppose, that any traitor will have his eyes opened to see the true character of the awful work in which he has been engaged, though it seems as if such an atrocious butchery were enough to make him see it; but of this be sure, that all loyal men are united now: and woe be to the secessionist who does not instantly sue for mercy or fly the country! I have seen them launch a great ship. The ways are laid solid and secure; and then the workmen split away, one after another, the blocks from underneath the keel. Gradually the huge structure settles upon the slippery ways, and glides majestically into her future element. The two ways under our ship of State are *justice* and *mercy*. In the providence of God, block after block has been knocked away, prop after prop removed, till now, just ready to glide into the new future, she is settling all her weight upon her ways,—ways made slippery by the blood of the murdered Chief Magistrate, and Minister: *woe, woe, woe* to him who puts himself in the line of her course! Infinitely better for him had he been strangled at the birth!

Be sure, this people will mourn from sea to sea; but be sure also, that any provocation will bring out the indignant, instant, sympathetic cry from every lip. "Die, traitors, assassins all! live, the republic, liberty, and law!"

The God of infinite justice and mercy be our helper! Amen.

NOTE. — Preached Sunday morning, April 16, after the news of the assassination of President Lincoln.

MEMORIAL SERMONS.

THE CAPTURE OF RICHMOND.

SOME OF THE RESULTS OF THE WAR.

THE ASSASSINATION OF THE PRESIDENT.

BY

EDWIN B. WEBB,

PASTOR OF SHAWMUT CHURCH, BOSTON.

◄ ● ►

BOSTON:

PRESS OF GEO. C. RAND & AVERY, 3 CORNHILL.

1865.

www.ingramcontent.com/pod-product-compliance
Lightning Source LLC
Chambersburg PA
CBHW022006190326
41519CB00010B/1409